Mites and Lice on Poultry

by US Dept. of Agriculture

with an introduction by Jackson Chambers

This work contains material that was originally published in 1917.

This publication is within the Public Domain.

This edition is reprinted for educational purposes and in accordance with all applicable Federal Laws.

Introduction Copyright 2017 by Jackson Chambers

IMPORTANT NOTE & DISCLAIMER

IMPORTANT NOTE :
As with all reprinted books of this age that are intended to perfectly reproduce the original edition, considerable pains and effort had to be undertaken to correct fading and sometimes outright damage to existing proofs of this title. At times, this task can be quite monumental, requiring an almost total rebuilding of some pages from digital proofs of multiple copies. Despite this, imperfections still sometimes exist in the final proof and may detract slightly from the visual appearance of the text.

DISCLAIMER :
Due to the age of this book, some methods or practices may have been deemed unsafe or unacceptable in the interim years. In utilizing the information herein, you do so at your own risk. We republish antiquarian books with no judgment or revisionism, solely for their historical and cultural importance, and for educational purposes.

Self Reliance Books

Get more historic titles on animal and stock breeding, gardening and old fashioned skills by visiting us at:

http://selfreliancebooks.blogspot.com/

INTRODUCTION

I am very pleased to present to you another important publication from the *U.S. Department of Agriculture*. **Mites and Lice on Poultry**, first published in 1917, is also known as ***Farmers' Bulletin No. 801***.

Although nobody is thrilled to read books on creepy crawlies and blood-sucking fiends that want to vampirize your flock, the knowledge in this book is essential to all poultry breeders and egg producers to maintain healthy birds.

No matter what you raise – egg layers, broilers, or even pure-bred show birds, keeping them free from parasites is essential for the health and quality of your stock.

The USDA books are always informative and knowledgeable, but keep in mind some of them are antiquated and may contain outdated and defunct practices, but they always contain sound, timeless advice also.

Jackson Chambers

State of Jefferson, November 2017

ASIDE from the chicken tick and the sticktight flea, the most important external parasites of fowls are the common red mite, the scaly-leg mite, and various lice.

The common mite sucks blood from the fowls and breeds in the cracks of the roosts and buildings. It may be destroyed by two or three applications of crude petroleum or certain coal-tar products to the roosts and buildings.

Scaly leg is caused by a small mite which may be destroyed by dipping the legs in crude petroleum.

In addition to a description of mites and lice this bulletin tells of a new but cheap and effective insecticide for use in destroying poultry lice. It is sodium fluorid, a white powder, which can be obtained through druggists. A single application, which costs about half a cent, will destroy all of the lice on a bird. Hundreds of fowls have been treated in the experiments conducted, but no injury whatever to them has occurred. Full instructions regarding methods of application are given in the bulletin.

MITES AND LICE ON POULTRY.

CONTENTS.

	Page.		Page.
Mites	3	Lice—Continued.	
The common chicken mite	3	Pigeon lice	19
Scaly-leg mite, depluming mite, and other mites	10	Lice of the guinea fowl and peafowl	19
		Control of poultry lice	19
Chiggers ("red bugs" or harvest mites)	11	Sodium fluorid effective against all lice	20
Lice	12	Other remedies for lice	24
Lice on chickens	12	Supplemental control measures for all pests	26
Lice on turkeys	17	A method of avoiding poultry pests	26
Lice on geese and ducks	18		

External parasites are one of the most important factors operating to retard the development of the poultry industry, but it is difficult to determine which of the parasites are of greatest importance. Both lice and mites are found in practically every locality where poultry are raised. Where present in any considerable numbers both lice and mites reduce egg production and hinder the growth and reduce the quality of flesh of all classes of poultry.

MITES.

THE COMMON CHICKEN MITE.[1]

Poultry raisers are all too familiar with the common red or gray mite which infests poultry houses. In general those who are making a specialty of poultry raising have comparatively little trouble with mites, or at least they keep them reduced to a point where they are of little importance. On the other hand, farmers and others who raise poultry as an incident to other operations frequently find their chicken houses overrun by mites. The attack of this blood-sucking mite is of an insidious nature which does not readily draw attention to its presence, and often the poultryman is not aware of an infestation until he is attracted to it by the irritation produced by mites on his own body through coming in contact with the infested coops. The presence of the pest may be determined readily by the detection of small areas on the boards specked with black and white as though dusted with salt and pepper. This is the excrement of the mites, which are hidden in adjacent cracks or rough places. More careful examination will reveal masses of mites in hiding, together with their eggs and the silvery skins cast by the young. In moderately infested

---

[1] *Dermanyssus gallinae* De Geer.

poultry houses the injury to the fowls is not very apparent, but the constant blood loss and irritation are shown by decreased egg production and the poor condition of the flesh of fowls. In heavily infested coops it is not unusual for the chickens to become droopy and weak, with pale comb and wattles. Sitting hens desert their nests and thus ruin the eggs or, as is often the case, they are found dead on the nest, being killed outright by the attack of thousands of mites. In extreme cases a considerable percentage of the fowls succumb, even though not sitting, and all are so weakened as to be very susceptible to various diseases.

DISTRIBUTION AND ABUNDANCE.

While the species sometimes becomes very numerous in the chicken houses in the northern part of the United States, the shorter breeding season there usually makes it of less importance than in the South where breeding continues throughout the year with little or no interruption. Although many assert that dampness has much to do with the abundance of the chicken mite, experience has shown that it occurs in rather greater numbers in the semiarid and arid regions of the Southwest than in the more humid parts of the South.

LIFE HISTORY AND HABITS.

Blood is absolutely essential for the development of this mite in all stages. The mite feeds almost entirely at night, except that it often feeds on hens on nests.

Chickens may carry a few mites (sometimes a hundred or more) in their feathers during the day following a night spent in infested quarters, but most of these leave the host during the following night. In some cases mites may remain on chickens during three days and nights, but nearly all become engorged and leave them by the third night.

Within 12 to 48 hours after receiving a meal of blood the mature female deposits from three to seven pearly white and elliptical eggs (see fig. 1), laid singly in the cracks in which the adults are hiding. The operation of feeding and depositing is repeated as many as eight times, and from 25 to 35 eggs in all are deposited.

In summer the eggs hatch in about two days, and one to two days later, without feeding, the larvae shed their skins and become nymphs (see fig. 2). With a very short rest these light-colored nymphs engorge with blood, secrete themselves, and molt their skins the second time 30 to 48 hours after having fed. These mites of the second nymphal stage soon engorge again, shedding their skins one to two days later and becoming adults. The grayish-colored unfed adult is shown in figure 3, and the engorged female, dark red in color and quite plump, in figure 4.

Thus the chicken mite reproduces very rapidly, the complete life cycle from egg to adult requiring not more than seven days.

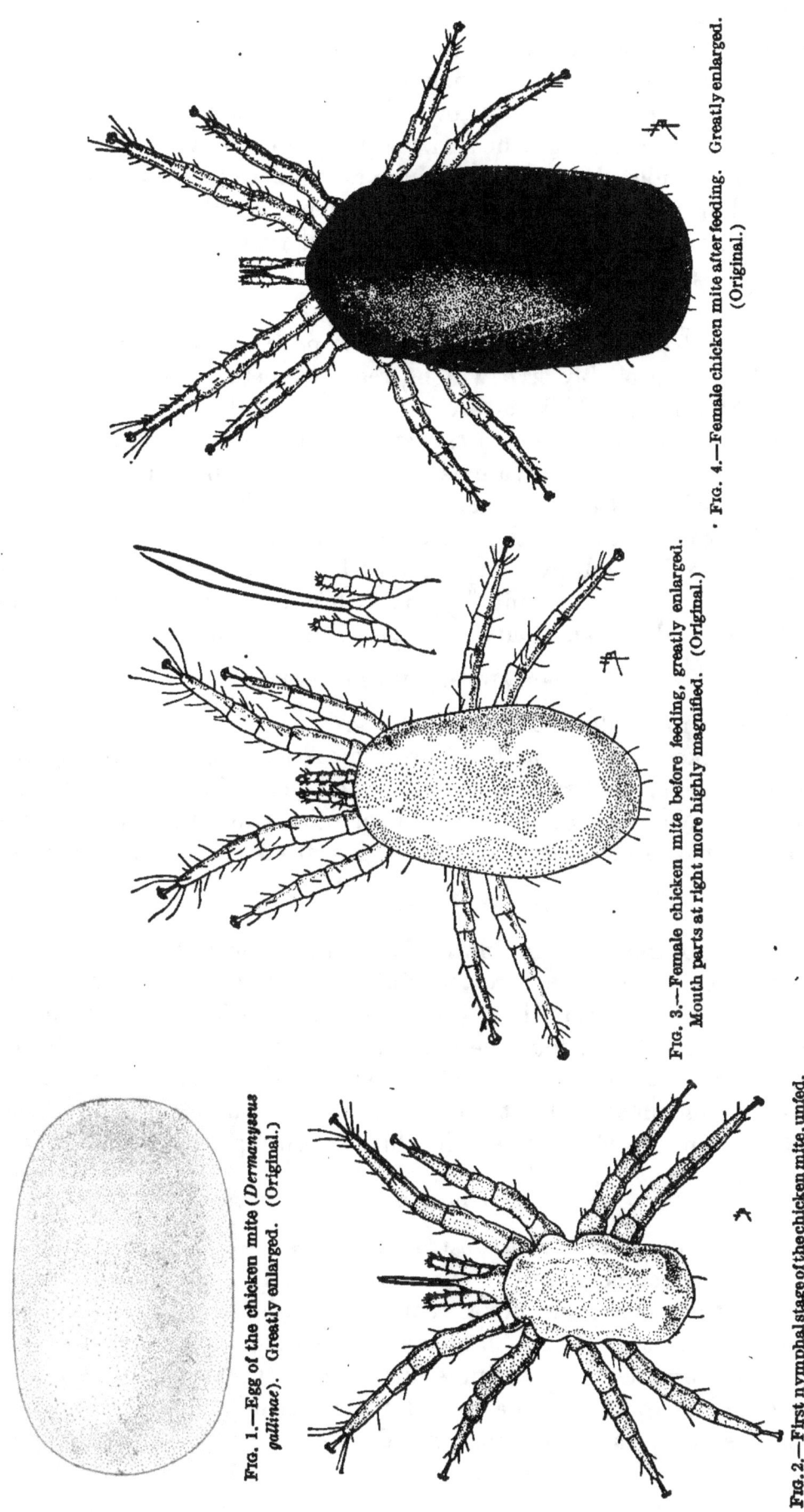

Fig. 1.—Egg of the chicken mite (*Dermanyssus gallinae*). Greatly enlarged. (Original.)

Fig. 2.—First nymphal stage of the chicken mite, unfed. Greatly enlarged. (Original.)

Fig. 3.—Female chicken mite before feeding, greatly enlarged. Mouth parts at right more highly magnified. (Original.)

Fig. 4.—Female chicken mite after feeding. Greatly enlarged. (Original.)

The weather is never too hot for this mite to thrive, and development is most rapid in midsummer. In the Southern States the mites are not entirely dormant during the winter, but feed and develop when the temperature is not low. This is also true in the North in chicken houses that are heated. Where some development takes place throughout the year, and where a complete generation of mites is developed in a week's time, hordes of mites will be present in a poultry house within a comparatively short time if something is not done to destroy them.

LENGTH OF LIFE.

It is probable that in a poultry house once infested at least four months and probably five will be required before all of the mites will starve if the chickens are removed from the house. In tests made by the writers some mites were still alive after a period of 113 days, and since these individuals were collected from an infested house it is not unlikely that they had matured some time previously. The tests indicate that where the mites are supplied with a certain amount of moisture they will live longer than when kept under very dry conditions. This may account, in part at least, for the idea that mites are worse in damp and badly ventilated chicken houses.

HOSTS AND METHODS OF SPREAD.

Chicken mites do not feed to any great extent upon other hosts when chickens are at hand. They are carried about chiefly by the interchange of poultry and in crates and boxes in which fowls are shipped. No doubt clean premises sometimes are infested by mites carried on the clothing of people going from one chicken yard to another.

CONTROL.

Owing to the fact that mites feed during the night and secrete themselves in cracks and crevices during the day, their presence very often is overlooked until a very heavy infestation has developed. In such cases they should be attacked energetically. Although not hard to kill, the greatest obstacle is the difficulty of reaching them in their hiding places. Dust baths will not control them, as at most only the few which remain on the chickens during the daytime will be destroyed.

TREATMENT OF INFESTED CHICKEN HOUSES.

The first step necessary to destroy the mites is to get rid of the hiding places so far as possible. The roosts should be taken down and all unnecessary boards and boxes removed. In heavily infested houses the mites are to be found in all parts of the building, including the roof. Where they are less numerous the infestations usually are confined to the roosts and nests and the walls immediately adjacent. For small coops a hand atomizer will suffice for applying insecticides as sprays, but for larger houses a bucket pump, knapsack sprayer, or barrel pump is desirable. A rather coarse spray should be applied from all angles and thoroughly driven into the cracks. The floor also should be treated, as many mites fall to the floor when the roosts are being removed.

In tests conducted during the last two years a considerable number of materials used as sprays have proved effective. One of the so-called wood preservers [1] was found immediately effective, and its killing or repelling power lasts for months. As this material is rather expensive (about $1 per gallon), and is too heavy to spray well, it is advisable to reduce it with equal parts of kerosene.

Crude petroleum is almost as effective, retains its killing power for several weeks, and in most localities it is very cheap. It will spray better if thinned with one part of kerosene to four parts of crude oil.

Both of these materials often contain foreign particles which should be strained out before the spraying is begun. It has been found that one thorough application of either of these materials will completely eradicate the mites from an infested chicken house, but ordinarily it is advisable to make a second application a month after the first, and in some cases a third treatment is required. These subsequent applications may be made with a brush, using the materials pure and covering only the roosts, their supports, the walls adjoining, and the nests if they are infested. This method of application is effective for the first treatment also if the houses are not heavily infested. Poultry should be kept out of the treated buildings until the material is well dried into the wood.

Pure kerosene and kerosene emulsion in double the strength ordinarily applied to plants will destroy all mites hit, but these substances have not body enough to destroy those mites which are in more protected situations, and several applications at 10-day intervals are needed to destroy all the mites.

Arsenical dip, such as is used to destroy cattle ticks, has been found fairly satisfactory for use against chicken mites. Several applications are required to eradicate the mites from poultry houses. In regions where cattle dipping is practised and this solution is readily available, it is perhaps the most convenient and cheapest material to use. Of course due care should be taken to avoid the accidental poisoning of the fowls. The standard coal-tar stock dips, used in solutions slightly stronger than are recommended on the cans, will destroy all mites reached by the spray, and in addition their germ-destroying properties are a desirable feature.

---

[1] The product referred to is employed extensively for the preservation of posts and other timbers set in the ground, and consists of certain coal-tar products known as anthracene oil, with zinc chlorid added. Crude carbolic acid is quite effective, but has less body and hence does not last long.

In tests made by the writers lime-sulphur solutions such as are used against scale insects proved much less effective than the insecticides already mentioned. Standard indoor whitewash [1] with 5 per cent of crude carbolic acid or cresol added gives good results, although not equal to those obtained by the use of crude oil or the wood preserver mentioned. Dry sulphur or lime will not control this mite.

With any insecticide the results will depend largely on the thoroughness of the application.

ROOSTS AND NESTS.

After the first spraying it is advisable to put in new roosts if the old ones furnish many hiding places for mites. The roosts should never be nailed to the side of the building but arranged so as to be easily removable. A convenient form of roost is shown in figure 5. The supports for the roost poles should consist of two 2 by 4's on edge in a horizontal position. The ends of these rest in notches cut in the ends of four uprights made of 2 by 6's and driven into the ground or nailed to the floor. The roosts should consist of smooth 1 by 3's or 2 by 2's, the ends resting in notches cut in the 2 by 4's. If the notches fit the poles closely it is unnecessary to nail the latter. The roosts thus are removed easily when the chicken house is to be cleaned, and a coat of one of the mite destroyers mentioned can be applied to the ends of the roosts occasionally.

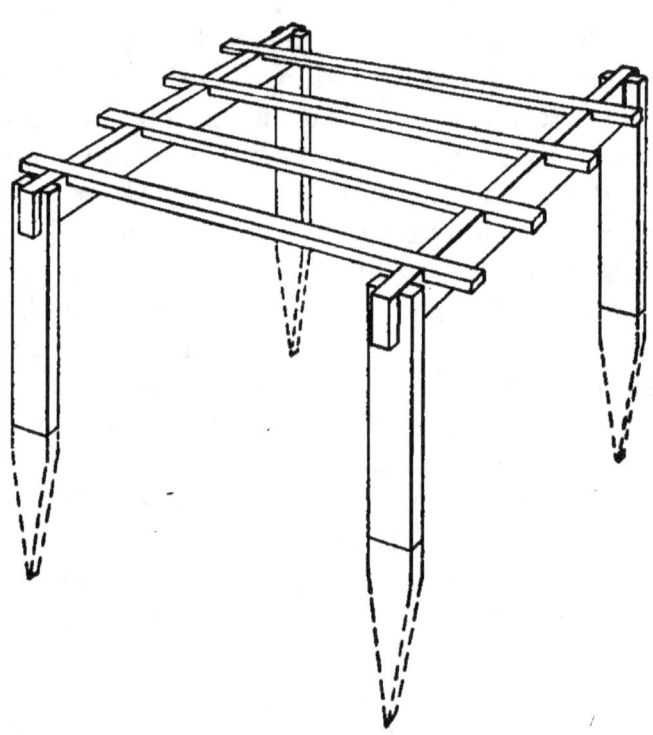

FIG. 5.—Chicken roost, suggesting method of making treatment for mites easy. (Original.)

If dropping boards are used they can be made to fit up to the four posts. In larger houses the horizontal 2 by 4's may be fastened to the back wall with hooks or certain types of screen hangers.

Another method of constructing the roosts, which is especially applicable to the Southwestern States where the chicken tick occurs, is to suspend a frame from the ceiling on baling wire and place the

---

[1] Methods of making whitewashes are discussed in Farmers' Bulletin 474 of the Department of Agriculture.

roosts across this frame. None of the structure should be allowed to come in contact with the walls, and there is then little opportunity for mites to reach the chickens. The underside of the roosts must be watched, however, to see that mites have not been introduced accidentally, as they have been known to breed on such roosts until present in considerable numbers.

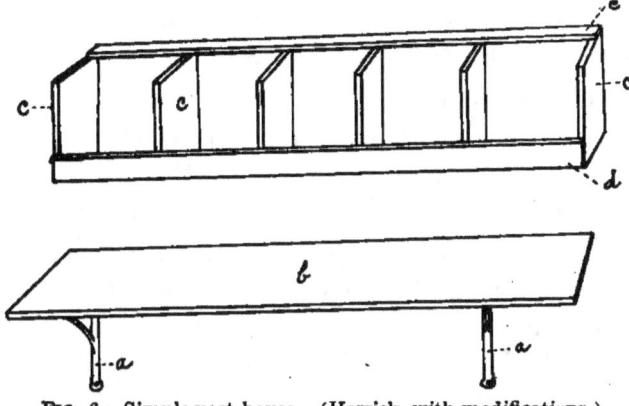

FIG. 6.—Simple nest boxes. (Herrick, with modifications.)

If convenient, the nests should be entirely apart from the roosting quarters. They may consist of boxes, which are easily handled, cleaned, or, if infested, destroyed. A series of nests made of boards is not objectionable if placed on a framework free from the walls of the henhouse and easily removable for cleaning. The simple arrangement devised by Prof. Herrick and illustrated in figure 6 may be used. Wooden or iron brackets (a) are fastened to the wall and upon these is laid a 12-inch board (b) which forms the bottom of the nests. The back of the nests is formed by the wall, and the partitions are made by cutting a 12-inch board into pieces 12 inches long (c) held upright by a 1 by 3 (e) nailed on top even with the back edges and a similar strip (d) nailed along the front at the bottom. The partitions and the bottom can be readily lifted off and thoroughly cleaned and the wall behind treated.

Great care should be taken to keep nests occupied by sitting hens free from mites. It is hard to work effectively against the mites when many hens are brooding; moreover, oil used freely about the house at that time may soil the eggs and prevent successful hatching. Infested quarters, therefore, should be treated thoroughly in the late winter before hens are set, so as to start them in nests which are absolutely clean. Beneath the straw of the nest a layer of lime and sulphur will tend to prevent mite breeding, and the entire nest may be dusted occasionally with pyrethrum. Broken eggs and the straw soiled by them should be removed promptly, as they tend to attract mites.

When poultry are to be transferred to new quarters it is desirable that they be kept three days and nights in a pen so that the mites will leave them before their introduction into the new building. The roosts in the new building and in the quarantine cage should be treated in order that any mites which have left the fowls may be destroyed.

SCALY-LEG MITE, DEPLUMING MITE, AND OTHER MITES.

Two species of itch mites attack fowls, one of which [1] is the cause of scaly leg. While this mite commonly remains on the feet, burrowing through the scales and causing their enlargement, it also attacks the comb and the neck. A crust of loose tissue is formed above the burrows, and intense itching results from this mining habit. When scaly leg is left untreated the feet often become badly distorted, and in some cases the fowl can scarcely walk or get up to the perch. Sometimes terminal joints of the toes are lost. As the mites are transferred from one bird to another, scaly-leg fowls should be treated promptly and should not be introduced among clean birds. Wood preserver or crude petroleum used on the roosts doubtless will aid in preventing the spread of the scaly-leg mite from one fowl to another. Applying crude petroleum to the legs with a brush or dipping the legs into this oil is very effective. One application usually is sufficient, but if the scales are not largely shed off after a lapse of 30 days the treatment may be repeated. Kerosene oil is applied by some farmers in the same way, but is less effective than crude oil. In using either, care should be taken not to get the oil on the upper part of the leg or on the feathers. A less severe but more laborious treatment consists of soaking the feet in warm soapsuds until the scales are loosened and then greasing the feet and legs with sulphur and lard, or lard containing 6 per cent crude carbolic acid.

The other itch mite,[2] commonly called the depluming mite, is a very small creature which burrows into the skin near the base of the feathers. The intense itching sometimes causes the fowls to pull their feathers until they are almost naked. Repeated applications of sulphur ointment should destroy these mites.

There are two other species of small soft-bodied mites sometimes found on poultry. One of these [3] bores into the skin. The other,[4] which has been found in several places in this country, occurs in the air passages, lungs, liver, and other internal organs of chickens and turkeys. Serious injury probably is not caused by these mites except when they are present in large numbers, when breathing may be hindered. Another small mite [5] sometimes feeds on the feathers of fowls but causes no apparent injury. Still another species [6] has recently been found by the writers in great numbers along the grooves on the under side of the shaft of the wing feathers of turkeys in Texas and Louisiana. Associated with this, but apparently in very small numbers, another mite [7] was taken. Neither of these caused any apparent injury to the host. Several other kinds of mites are found on various birds, as well as domestic fowls, but these are of little or no importance as parasites.

---

[1] Known scientifically as *Cnemidocoptes mutans* Robin. [2] *Cnemidocoptes gallinae* Railliet. [3] *Laminosioptes cysticola* Vizioli. [4] *Cytolcichus nudus* Vizioli. [5] *Rivoltasia bifurcata* Rivolta. [6] *Freyana chanayi* Trouessart. [7] *Megninia cubitalis* Megnin.

CHIGGERS ("RED BUGS" OR HARVEST MITES).

The chiggers which attack chickens are the same minute red mites which attack man. They are the first stage of a large red mite, which when mature is entirely harmless. Normally these immature mites are parasitic upon insects. They are often very widely distributed in fields and thus readily picked up by chickens. They attach themselves to the skin in groups beneath the wings and on the breast and neck. The injury is most severe among young chickens, although grown fowls occasionally are annoyed to some extent. Young chickens which have a free range, especially if it extends into lowlands and under trees, are very susceptible to attack. The infested chickens become droopy, emaciated, soon refuse to eat, and if exposure to the mites is continued a considerable mortality is likely to result. Intense irritation is set up, and abscesses are formed at the points where the clusters of mites are feeding. These abscesses sometimes are one-third of an inch in diameter and surrounded by a greater inflamed area. Suppuration takes place beneath the skin, and swelling around the clusters of mites causes the formation of a considerable cavity at the center where the mites are attached.

In the South and in the Central States, where chiggers are numerous, probably the best plan is to keep young chickens during the summer from ranging where these mites are likely to occur. If chickens are hatched very early in the spring it is likely that they will escape chiggers more or less completely. When the chickens do become infested the application of sulphur ointment or kerosene and lard will destroy them. If extensive suppuration has taken place the scab should be removed and the area washed with a 4 per cent carbolic-acid solution. Occasional light dusting of chickens with flowers of sulphur doubtless will keep these "red bugs" off, and where fenced range is infested the application of sulphur at the rate of 50 pounds per acre with a dust blower would keep them in control.[1]

LICE.

LICE ON CHICKENS.

All poultry lice or bird lice have stout cutting or biting mouth parts which distinguish them from the sucking lice of cattle and other domestic animals. Unlike the mites, lice remain on the hosts constantly. More than 40 species of lice are found on the various domestic fowls. Some species are found on one host only, while other kinds may attack a number of fowls. Chickens are infested by more kinds of lice than any other domestic fowl. Seven species are very commonly found on chickens in the United States, four or five on

---

[1] Further information regarding harvest mites or red bugs may be had from Farmers' Bulletin 671, which may be obtained on application to the Secretary of Agriculture, Washington, D. C.

pigeons, two or three each on geese and ducks, three on turkeys, and several each on guinea fowl and peafowl.

All these lice are adapted to the conditions under which they live. They have a flattened form and are fitted with various spines and peculiarly modified legs which assist them in moving about through the feathers. Certain species which remain on the larger feathers have a very narrow, elongate form which utilizes the protection afforded by the grooves between the barbs of the feathers. In fact, poultry lice show a wide divergence in size, shape, and spiny armature.

FOOD HABITS AND INJURIOUSNESS.

Poultry lice are not fitted for sucking blood. They feed on portions of the feathers or on scales from the skin, and their presence in any considerable numbers is responsible for serious injury.

In the Southern States the loss due to lice probably is greatest among young chickens. Chickens hatched after April 1 and brooded by hens experience a high mortality, much of which appears to be due directly or indirectly to lice. Early chickens also are sometimes affected. The lice often leave the hens and pass to the chickens before these become dry after emerging from the shell.

The first symptoms of lice infestation usually are droopiness, lowered wings, and ruffled feathers. Diarrhea follows, and the chickens then often die in a few days, or, when older, sometimes fall a prey to various diseases. Grown fowls sometimes may be very heavily infested with lice without showing any ill effects, but in such cases the egg yield is likely to decrease. In other cases the fowls may lose weight and sometimes die as a result of the lice or succumb to some of the common chicken maladies.

Turkeys suffer to a considerable extent when young, and no doubt poults frequently are killed by gross infestation. Older birds do not seem to be so badly affected. This is also true of ducks and geese. In general, these fowls are less heavily infested with lice than chickens.

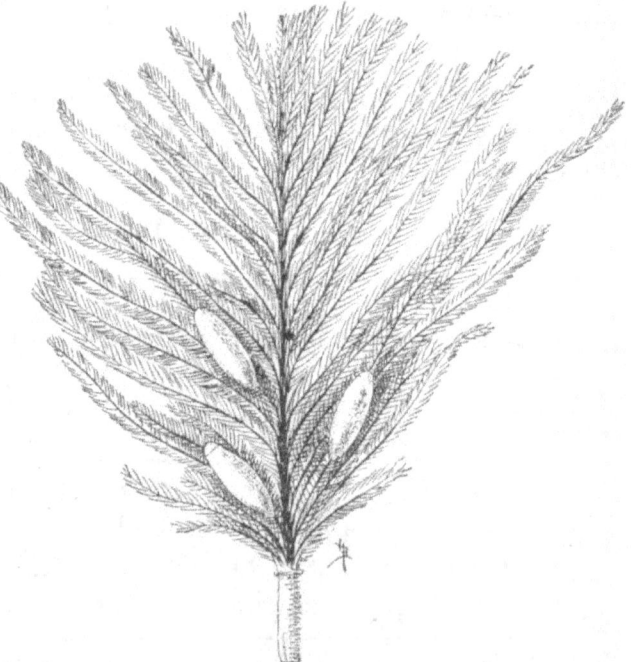

FIG. 7.—Eggs of the head louse (*Lipeurus heterographus*) on feather. Greatly enlarged. (Original.)

KINDS OF LICE ON CHICKENS.

The seven different species of lice common on hens are spoken of as body lice, head lice, and feather lice, according to the usual places in which they are found, but since the different species intermingle to a considerable extent, it is not possible to separate them absolutely on this basis. The writers have observed that the relative number of lice of the different species varies much in different flocks in the same neighborhood, and even in the same flock some chickens often have one species predominating, while others have another. Usually three or more species are to be found on an infested fowl.

THE HEAD LOUSE OF CHICKENS.[1]

This species is primarily a head louse, although occasionally found on the neck and elsewhere. It is undoubtedly the most injurious species to young chickens, as many of the other forms which are serious annoyers of grown poultry do not thrive well in the down on chicks. It is a dark grayish species nearly one-tenth of an inch in length, and may be found on the top or back of the head, behind the ears, or beneath the bill. Usually it is located close to the skin with its head very close to or against the skin of the chicken, the body extending away from the skin on the down or along the feathers. The eggs are deposited singly on the down or small feathers about the head. Eggs attached to a small feather are shown in figure 7. These hatch in four or five days into minute semitransparent lice which resemble the adult in shape. After molting the skin several times, and in the meantime increasing in size and becoming darker in color, the lice reach the adult stage in about 17 to 20 days. The male of this species is shown in figure 8.

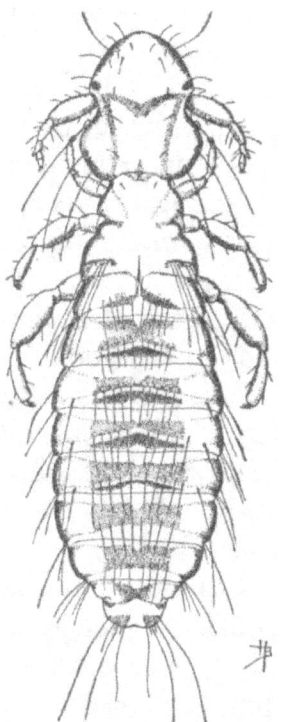

FIG. 8.—Head louse: Male, top view. Greatly enlarged. (Original.)

Despite the fact that this louse confines its attack principally to the head, it passes readily from one chicken to another and from the mother to her young. This is the species against which the poultryman must guard his young chickens. The treatment recommended for various lice (p. 22) is satisfactory for this species. It is essential that the applications be made to the regions about the head to destroy it on grown fowls, and on young chickens this is the only region which requires attention. After young chicks are fairly well feathered the head louse decreases in number, probably because conditions are less

---

[1] *Lipeurus heterographus* Nitzsch.

favorable for breeding and because the older chickens scratch the infested parts more vigorously. The number of head lice may increase again after the chickens become adult.

THE BODY LOUSE OF CHICKENS.[1]

The common name "body louse" is aptly applied to this species, and refers to its habit of remaining on the skin of the fowl rather than on the feathers. It does not always confine itself to the body, sometimes being taken on the head, neck and legs. It favors those portions of the skin which are not densely feathered. On chickens it is partial to the region just below the vent, but in heavy infestations it is abundant on the breast, under the wings, on the back, and also on the head, neck, and thighs. When the feathers are parted it is seen running rapidly upon the skin to seek protection. With young chickens it is more abundant on the back than around the vent.

This louse is rather large and robust, straw yellow in color, with some dark spots due to food within the digestive tract. The two sexes are shown in figures 9 and 10.

The body louse is probably the most injurious species on grown chickens, but it also infests young fowls, sometimes seriously. As it remains on the skin of the host, irritation is kept up constantly. Often a marked reddening of the skin of the fowl in the regions most heavily infested results, and in some cases scabs and blood clots are formed.

The eggs are deposited in clusters on the base of the feathers, usually being attached to the lowest barbs along the shaft. They are most abundant on the small feathers below the vent, where the masses of eggs sometimes become very large—fully half an inch in length. As the lice continually add eggs, the masses are extremely large when seen several months after molting. In the case of young fowls the eggs are often deposited in numbers on down or small feathers and on hairs about the head and throat. A mass of eggs of this species is shown in figure 11.

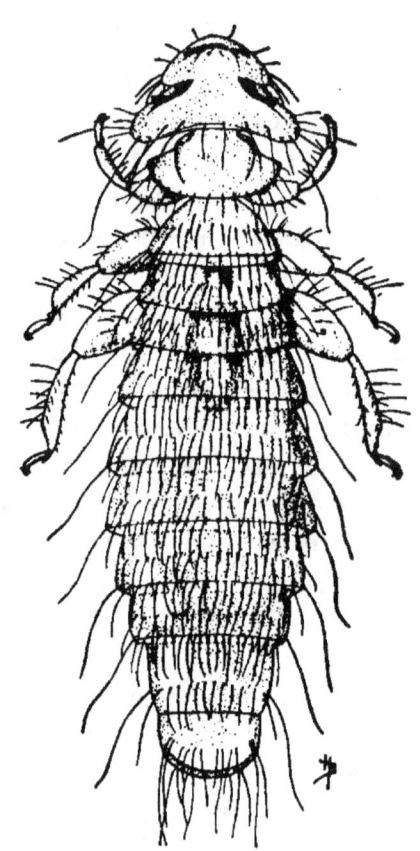

Fig. 9.—Body louse (*Menopon biseriatum*): Male, top view. Greatly enlarged. (Original.)

---

[1] *Menopon biseriatum* Piaget.

The eggs hatch in about a week, and the adult stage is reached from 17 to 20 days after the eggs are deposited. This louse has a short period of growth both in summer and in winter; hence fowls which are not actively fighting the louse become swarming with them in a very short time. Fortunately the heat of the body is necessary for the hatching of the eggs, and the lice themselves die in a very short time when off the fowl. For this reason little attention need be given to lice and eggs which are shed by the host during molting or at other times. This point, as well as methods of control, is discussed in later pages.

The body louse appears to pass readily from one fowl to another when they are closely associated. It also infests turkeys, upon which it multiplies to some extent, and it is said to occur on pigeons, but has not been found on them by the writers.

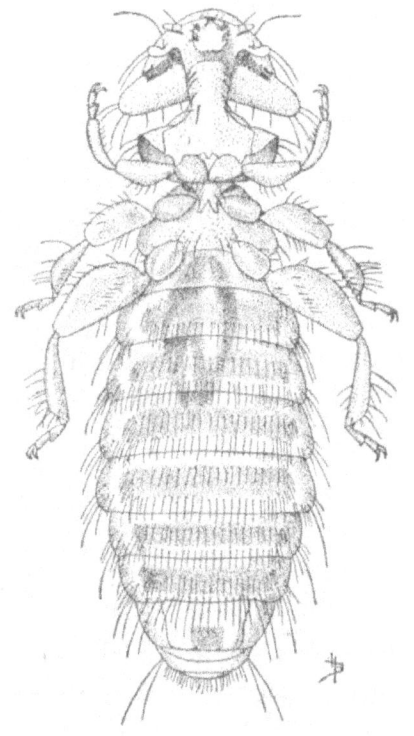

Fig. 10.—Body louse: Female, underside. Greatly enlarged. (Original.)

THE SHAFT LOUSE.[1]

The shaft louse is the species spoken of by most authors as the "small body louse," a name which does not fit the habits of the species. Normally, it occurs along the shaft of the feathers and does not remain on the body of the host for any length of time. The shaft louse is closely related to the large body louse and resembles it somewhat. It is smaller, rather lighter yellow in color, and somewhat less spiny. (See fig. 12.) The habits of the shaft louse will enable one to separate it readily from the large body louse. When the feathers on the back or breast are parted this louse will be seen running toward the body along the shaft of the feather. Sometimes as many as a dozen lice will be seen, one behind another, along the feather shaft.

Although this species is probably the most common found on chickens in various parts of the country, the writers consider it of much less importance than the body louse, chiefly because it stays on the feathers the greater part of the time and probably feeds exclusively on the barbs of the feathers and on scales along the shaft. It is not known to occur on young chickens. Seemingly the absence of feathers prevents the successful development of the species on young fowls.

---

[1] *Menopon pallidum* Nitzsch.

The eggs are deposited singly at the base of the feathers, hidden between the main shaft and the after shaft. It appears that eggs of the shaft louse require more time for incubation than those of the head louse or the body louse, and the time required to reach maturity is also greater. The shaft louse appears to live much longer on feathers which have dropped from the host than any other species on domestic fowls.

Several kinds of domestic fowls harbor the shaft louse, but it has not been shown that they will breed successfully on fowls other than the chicken. It has been found on the guinea fowl and on turkeys and ducks closely associated with chickens.

FIG. 11.—Mass of body louse eggs attached to feather. Greatly enlarged. (Original.)

THE WING LOUSE.[1]

This species has been called the "variable louse," but the variations are not apparent to the ordinary observer, and the writers suggest "wing louse" as a common name. This is the only species found commonly on the large wing feathers of chickens. It is seen at times also on the neck hackles, tail, and back feathers.

The wing louse, which is related to the head louse, is dark gray and has an elongate body. It is more slender than the head louse, however, and rather darker in color. Most easily seen on white fowls, it is found in all situations, but especially along the underside of the primary wing feathers. It is a sluggish species and often lies between the barbules of the feathers near the shaft without showing any life. The elongate white eggs are laid between the barbules of the large feathers.

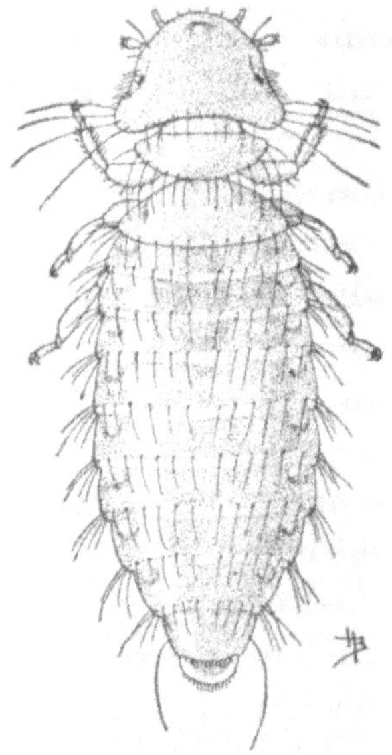

FIG. 12.—Shaft louse (*Menopon pallidum*): Female, top view. Greatly enlarged. (Original.)

OTHER LICE OF CHICKENS.

Three other species of lice are found more or less commonly on chickens. The species to which the writers have applied the common name of "fluff louse"[2] is very small but broad,

---

[1] *Lipeurus variabilis* Nitzsch. [2] *Goniocotes hologaster* Nitzsch.

pale in color, with a translucent appearance. It is common on fowls, but seldom abundant, and is of little importance. It is found on the fluff of the feathers on various parts of the bird, but is most abundant where the feathers are fluffiest. Usually it hangs to the loose barbs on these feathers some distance from the shaft and shows little activity.

The large hen louse [1] is less abundant than the fluff louse. When present, it is easily recognized by its very large size and striking appearance. It is nearly an eighth of an inch in length and very broad in proportion, as shown in figure 13. The color is smoky gray to almost black, with darker marks on the sides of the abdomen. It occurs on the feathers on various parts of the chicken's body and is remarkably agile for its size. Some call it the "blue bug," hence it has become confused in certain instances with the chicken tick, for which the name blue bug is generally used.

The brown chicken louse [2] has not been reported to occur in America heretofore. The writers have taken it in several instances on chickens in the vicinity of Dallas, Tex., and Orlando, Fla. This indicates its presence in much of the South. It is somewhat smaller than the large hen louse and reddish brown in color. It is found on the feathers of the body. None of the three species last discussed has been taken by the authors on young chickens.

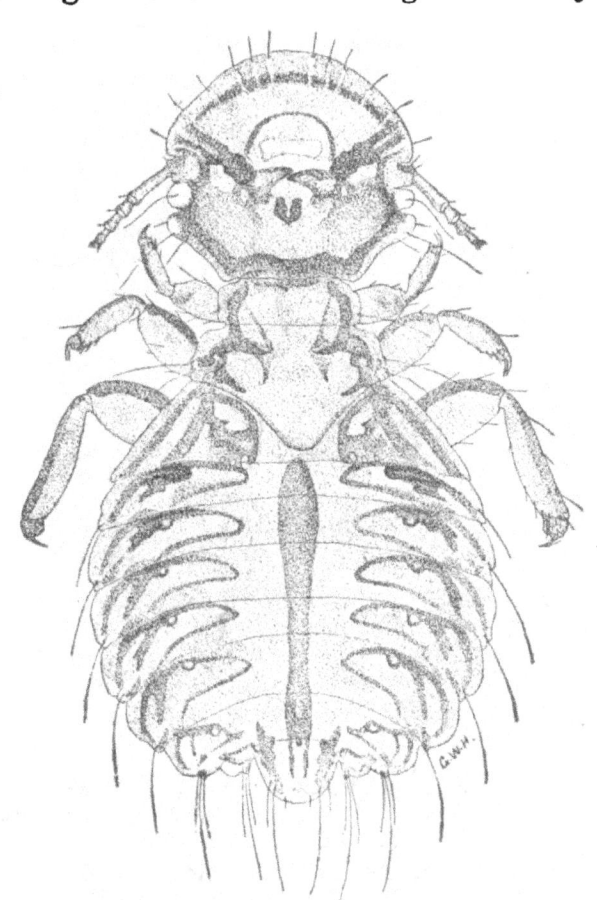

FIG. 13.—Large hen louse (*Goniocotes abdominalis*): Male, top view. Greatly enlarged. (Herrick.)

LICE ON TURKEYS.

Four species of lice are commonly found on turkeys in this country. One of these, which occurs particularly on turkeys associated with chickens, is the common body louse of chickens. The writers have not found this species in great numbers on turkeys, but it sometimes becomes sufficiently abundant to cause considerable irritation and doubtless is injurious both to the grown fowls and to the young.

---

[1] *Goniocotes abdominalis* Piaget.   [2] *Goniodes dissimilis* Nitzsch.

The shaft louse of chickens also has been found on turkeys, but probably does not breed on that host. The other two species seem to be native to the turkey, probably existing on this fowl in the wild state. The large turkey louse[1] (fig. 14) probably is most abundant. It occurs on the feathers on various parts of the body, especially on the neck and breast. The slender turkey louse is a species of good

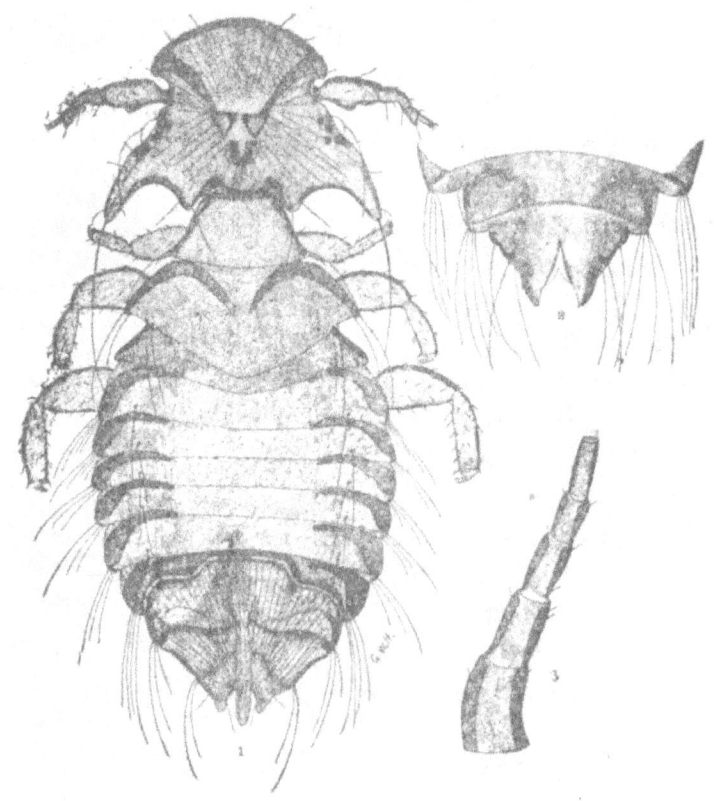

FIG. 14.—Large turkey louse (*Goniodes stylifer*): *1*, Male, top view; *2*, tip of abdomen of female; *3*, antenna of female. All greatly enlarged. (Herrick.)

size, though rather elongate, resembling in shape the head louse of chickens. Normally neither of these species is excessively abundant, but on crippled or unthrifty turkeys they may cause serious annoyance and undoubtedly they are injurious to poults.

For remedial measures see pages 19–26.

LICE ON GEESE AND DUCKS.

While considerable numbers of lice are found on domestic geese and ducks, they seldom become sufficiently numerous to cause noticeable injury. One of these species[2] is quite common on ducks throughout the country, and a variety of this same species is to be found on the goose. A slender species[3] has been found by the writers rather commonly on the duck in the vicinity of Dallas, Tex.,

---

[1] *Goniodes stylifer* Nitzsch.   [3] *Lipeurus temporalis* Nitzsch.
[2] Known scientifically as *Docophorus icterodes* Nitzsch.

and in other parts of the United States. Another louse similar in form, but which appears to be new to science, has been collected on these hosts in Mississippi. Both of these species inhabit the wing feathers and are often very numerous at the base of the large feathers of the wing. Young ducks which have been hatched by hens are sometimes much annoyed by the head louse, which attacks them in the same way as it does young chickens.

The same control measures recommended for chicken lice will destroy these insects on ducks and geese.

PIGEON LICE.

Considerable annoyance to pigeons has been reported by poultrymen in different parts of the country. While a number of species are reported from pigeons, most of the trouble seems to be due to the slender pigeon louse [1] and the broad pigeon louse.[2] The former sometimes occurs in great numbers, attacking both the old birds and the partially feathered squabs.

For measures of control see the discussion of this subject in subsequent pages.

LICE OF THE GUINEA FOWL AND PEAFOWL.

The guinea fowl and peafowl are both subject to the attack of several species of lice. Most of these are of kinds different from those found on the common chicken and other domestic fowls, but the guinea fowl has been found to become infested with four of the species found on the chicken. It does not appear that either of these hosts is materially injured by lice, but it is necessary to bear in mind the risk in allowing guinea fowl to remain untreated when eradicating lice from other poultry on the same farm.

CONTROL OF POULTRY LICE.

It has been generally felt that poultry lice are more or less a necessary evil and that the best that can be expected is to keep them in control by repeated treatment. A few have attempted, with varying degrees of success, to start with clean premises and clean fowls and keep them free from vermin. This is most feasible in the case of persons going into the poultry business on a rather extensive scale and with entirely new equipment (see page 26). For the average farmer and the poultryman already established the situation has resolved itself largely into a fight against the various pests already present.

There is no fundamental reason why a flock should not be entirely freed from lice and maintained in this condition. Reinfestation comes principally from stray fowls which gain access to the poultry yards and from purchased stock added to the flock. Stray fowls

---

[1] *Lipeurus baculus* Nitzsch.   [2] *Goniocotes compar* Nitzsch.

can not always be excluded, but in the case of added stock it is advisable to treat all chickens, old or young, when they are first brought on the premises.

The time of year for starting a campaign against lice is another point to be considered. The writers would favor the treatment of the entire flock during the late summer or early fall. At this time of the year weather conditions usually are favorable to dipping, most of the young fowls are well matured, and much of the superfluous stock has been disposed of, so that there are fewer birds to treat. Since there is very little danger of reinfestation from lice on molted feathers, the question of avoiding the molting period is not a serious one, yet, if the treatment can precede molting it probably would be better.

If the fall treatment has been neglected it is imperative that the flock be cleaned of lice before brooding time in the spring. Usually this would mean that the dusting method would have to be followed on account of adverse weather conditions. Treating the birds at this time will insure their vigor as well as undisturbed brooding, which is necessary to successful hatching, and, what is more important, the infestation of the young chickens will be avoided. Although lice normally stay upon the host continuously and do not have the habit of hiding away in cracks about buildings, yet the poultry houses and runs should be well disinfected occasionally, especially as action against mites is necessary if these are present. It is well to make this general clean-up at the time the flock is treated for lice. This minimizes any danger of reinfestation.

On large poultry farms the complete eradication of lice is often complicated by inability to control the fowls. When proper pen construction is at hand, it is possible to treat a pen or two a day until the entire flock is covered. The more rapidly treatment progresses the better, of course, and great care should be taken to avoid the escape of fowls from infested to uninfested pens.

SODIUM FLUORID EFFECTIVE AGAINST ALL LICE.

The writers have tested a number of the materials most generally advocated for lice destruction and several new compounds which it was thought might be effective. In this series of tests nothing else was found to be as satisfactory as sodium fluorid. The experiments have demonstrated that this chemical is exceedingly poisonous to all species of chicken lice. It kills both adults and young, including the young which emerge from the eggs present at the time of treatment.

Sodium fluorid can be obtained in two forms, known as commercial and as chemically pure. Both of these are in a dry state, the former being a dry powder and the latter consisting of small crystals somewhat lumpy. While the chemically pure material is effective, it is not as easily applied by the dusting method as the more finely powdered commercial form, and, furthermore, it is higher in price. The commercial grade should contain 90 to 98 per cent sodium fluorid.

This material is the sodium salt of the chemical element known as fluorin and hence is a compound very similar to ordinary table salt, which is known chemically as sodium chlorid. In asking for sodium fluorid it is therefore important that the name "fluorid" be carefully stated to the druggist. Up to this time the demand for the material has been very limited. Prior to the work of the writers its only known insecticidal use was against cockroaches, for which it has been demonstrated to be very effective. Owing to this limited demand the material is not ordinarily found in drug stores. Druggists, however, can obtain it readily from manufacturing chemists in the larger centers, and with demand it will be carried in stock by many local dealers. The commercial sodium fluorid at present prices should be retailed at from 30 to 60 cents per pound, the price varying somewhat with the amount ordered by the druggist and the distance from the chemical manufacturing centers.

Sodium fluorid in a dry state does not deteriorate quickly. It should be kept in a dry place either in bottles with stoppers or in closely covered cans. In this condition it will remain active indefinitely.

METHODS OF APPLICATION.

In treating poultry with sodium fluorid, *if proper methods are followed, a remarkable degree of control is obtained. One application of sodium fluorid to all fowls on given premises will completely destroy all lice present.* It is essential to make sure that the treatment is thorough and that every fowl is treated, for if one infested chicken escapes it will in a short time reinfest the entire flock and thus make it necessary to do the work over at a considerable loss of time and money.

Sodium fluorid may be applied in two forms, as a dust and as a dip. In using either form the first step is to see that all fowls are shut in the poultry house or placed in coops prior to beginning treatment.

DUSTING.

The action of sodium fluorid when applied in dust form is comparatively slow; hence, if fowls are examined the day following treatment, or even two or three days later, some lice may be found. The material persists, however, and after four or five days all lice disappear. Apparently the hatching of the eggs is not prevented, but the young lice find sufficient material present in the feathers upon emerging from the eggs to destroy all of them.

For complete destruction of lice it is essential to place small amounts of the material on different parts of the infested birds. Contrary to the usual belief, all species of lice do not migrate freely from one part of the bird to another, hence the material must be well distributed to bring it in contact with all lice present.

The writers have found what they term the "pinch method" to be entirely effective against all lice and to have the advantage of economy of time and material. When applying the material by this method (see illustration on title page) it is placed on a table in an open vessel, and the fowl is held by the legs or wings with one hand, while with the other hand a small pinch of the chemical is placed among the feathers next to the skin about as follows: One pinch on the head, one on the neck, two on the back, one on the breast, one below the vent, one on the tail, one on either thigh, and one scattered on the underside of each wing when spread. Each pinch can be distributed somewhat by pushing the thumb and fingers among the feathers as the material is released. It is advisable when dusting to hold the chicken over a large shallow pan, as in this way the small amount of material ordinarily lost is recovered.

The material may be applied by means of a shaker, but this method has some disadvantages as compared with the pinch method. Small nail holes are punched in the bottom of a can, which is provided with a close-fitting lid on the other end. The material is then shaken into the feathers with one hand, while the feathers are opened with the other. This necessitates the presence of a second person to hold and turn the fowl. When this method is followed the amount of sodium fluorid used may be reduced by adding four parts of some finely powdered material, such as road dust or flour, to each part of the chemical. If the material is employed alone, somewhat more of it is used than by the pinch method, and more or less dust floats in the air, which causes irritation of the throat and nose. This can be avoided largely if the operators wear dust guards over the nose or keep pieces of wet cloth over the nose and mouth.

Although the writers have not applied this material with a dusting machine or revolving barrel, they are of the opinion that this would not be thorough, might bruise the fowls, and would be irritating to the air passages of the birds.

For lice on young chickens, young turkeys, and in fact all newly hatched fowls the application of sodium fluorid in the dust form is recommended, rather than by dipping. This applies also to sick fowls. While the writers have not tried dipping against lice on pigeons, the dusting method has been found effective.

DIPPING.

There seems to be a general sentiment among poultry raisers against the practice of dipping fowls. This is probably partially on account of the fact that the dips tried have been of an oily or caustic nature and have tended to soil the feathers and in some cases injure the skin of the fowl and give the feathers a thorough wetting.

The experience of the writers does not justify this aversion when dipping in a sodium-fluorid solution. It may be said that in general the dipping method is most applicable to the Southern States and to summer treatments in the north. The first requisite is a rather warm sunny day so that the fowls will dry quickly. Windy weather should be avoided. In dipping fowls as described below, the feathers do not get thoroughly wet, and if the operation is finished an hour before sundown the fowls will become thoroughly dry before going to roost. In rather extensive tests of this method the writers have observed no ill effect whatever from the dipping. As compared with dusting, this method has an advantage in that it reduces considerably the cost of materials, is more rapidly done, and the discomfiture to the operator is avoided. It is just as effective as dusting.

The lice die much more quickly following dipping than when sodium fluorid is applied in dust form. It appears that all those which are touched by the liquid die very promptly, and the others succumb in a few hours.

In using the dipping method all that is necessary is a supply of tepid water and a tub. If two persons are to dip at the same time it is advisable to use a large tub. The water should be measured into the tub and three-fourths to 1 ounce of commercial or two-thirds of an ounce of chemically pure sodium fluorid added to each gallon of water. It is readily dissolved by stirring. The tub should be filled to within 6 or 8 inches of the top, and as the amount of solution is lowered through dipping numbers of fowls, water with the proper proportion of sodium fluorid dissolved should be added from time to time. In dipping the fowls it is best to hold the wings over the back with the left hand and quickly submerge the fowl in the solution, leaving the head out while the feathers are thoroughly ruffled with the other hand so as to allow the solution to penetrate to the skin on different parts of the bird. The head is then ducked once or twice, the bird is lifted out of the bath and allowed to drain a few seconds and is then released. The total time required for an individual fowl is from 30 to 45 seconds.

EFFECT OF SODIUM FLUORID ON FOWLS AND MAN.

Fortunately this compound is very destructive to lice without producing any ill effects on the chickens. No skin irritation or injury to the condition of the feathers has been observed in the large number of domestic fowls used in experimental work, when either the dusting or the dipping method was used. In dusting fowls there is occasionally some temporary irritation of the air passages, as evidenced by labored breathing and sneezing. This effect is not noticeable a few minutes after treatment.

Where some of the sodium fluorid in the dust form reaches the body of the operator and is allowed to remain for a number of hours, as may be the case in dusting several hundred fowls, local irritation and burning may occur on tender portions of the skin. In dusting large flocks it is therefore advisable to do the dusting on a table rather than to hold the fowls between the knees as is sometimes done. The solution does not injure the hands, even when dipping is continued for a number of hours, except in cases where sores are present which may become slightly irritated.

Precaution should be taken not to allow sodium-fluorid solution to remain in galvanized vessels any great length of time. In fact, it is best not to keep it over night in tubs or galvanized containers, as it will injure them.

COST OF APPLICATION.

One pound of commercial sodium fluorid, when applied by the "pinch method," will treat approximately 100 hens; thus at a cost of 40 cents per pound the expense for material will average less than one-half cent per fowl. It has been found by actual practice in treating several hundred fowls that an average of from two to three minutes is required for treating each fowl, one man doing the work. This includes the time necessary for catching the birds as well as dusting them. The dusting itself occupies about one to two minutes. Of course, the time involved in catching them would vary in every individual instance according to conditions. Using the above figures as a basis, and figuring a man's time at 20 cents per hour, it would cost approximately $1.25 to treat 100 fowls by the pinch method.

In treating with the dust can the amount of material is usually double and the average time per bird is somewhat increased.

By the dipping method the amount of the material is considerably reduced, especially if large flocks are to be treated at one time. Over 800 fowls have been dipped at one time, using on the average 5.2 ounces of sodium fluorid to 100 fowls, which at the same figure would cost 13 cents. The labor involved is also thus reduced. The average time for catching and dipping the birds was about one and three-fourths minutes per fowl, one man working. This makes a cost for labor, as above computed, of about 58 cents per hundred fowls and a total cost for material and labor of about 71 cents. This reduces the cost so that it is within the reach of every one, especially when it is considered that ordinarily much time is occupied in fighting lice without accomplishing that complete destruction which would result from a single treatment as above outlined.

OTHER REMEDIES FOR LICE.

While the use of sodium fluorid is advised in all cases, it may be stated that one application of flowers of sulphur when applied thoroughly in dust form has been found to destroy all stages of several species of lice experimented with. In a few instances, however, some lice remained on the fowls after treatment. This was attributed to the difficulty of getting the dust over every portion of the fowl, but at the same time it shows that exceedingly thorough and careful application of sulphur is required to secure complete destruction. About four days are required for the fowls to be freed of living lice. The ready availability of flowers of sulphur and its comparatively low cost per pound tend to recommend it for this use. Furthermore, it is not disagreeable to handle.

A number of poultry raisers, however, have stated that injury to the fowls sometimes results from the use of sulphur, although the writers have seen no injury further than a very slight scaling of the skin following treatment. It is possible that the injury observed by some was due to mixing the sulphur with grease or other substance. To accomplish complete destruction the writers have found it necessary to use considerable quantities of sulphur, averaging about 6 pounds per hundred fowls, which at 10 cents per pound would make a cost of 60 cents for the material. The expense of application would be about one-half greater than that given for the use of sodium fluorid by the pinch method, as it is necessary to apply the sulphur with a dust can. The total expense would therefore be greater than by using sodium fluorid by the pinch or the dipping method.

The writers have found also that dipping fowls in a soap solution made by dissolving 1 ounce of laundry soap in a gallon of water will destroy all lice present, but a second dipping 10 days later is necessary in order to destroy the lice that have hatched from eggs which are not killed by the treatment. This soap solution causes a complete wetting of the feathers, and hence there is no doubt danger of producing colds when the weather is unfavorable. It should be used only during favorable weather.

A great number of remedies are in general use in this country, only a few of which can be mentioned here. A mixture of crude carbolic acid, gasoline, and plaster of Paris is quite effective in reducing the number of lice, but experiments have shown that at least two and perhaps more applications are necessary to destroy all lice.

Mercurial ointment or blue ointment is also advised. It has been found that the use of this material as recommended will greatly reduce the number of body lice but has little effect on the head and wing lice. When several times the amount usually recommended is applied to a number of places on a fowl it is quite effective, but the cost of the material and treatment is greater than in the case of sodium fluorid, the use of greasy material is objectionable and burns result.

A number of other compounds, many of which contain pyrethrum, are advocated. These also fail to accomplish complete destruction of the lice. For head lice on young chickens carbolated petrolatum applied in small quantities has been found quite satisfactory. Medicated nest eggs, said to control poultry lice, are on the market. For the most part these consist largely of naphthalene. While this material will destroy lice when applied generally to the fowl, it is markedly injurious to the hen's eggs as well as to the bird. If used in quantity, or if the medicated eggs are allowed to remain for a considerable length of time beneath a hen, she may die as a result.

SUPPLEMENTAL CONTROL MEASURES FOR ALL PESTS.

Chickens will not give adequate returns in eggs or growth when kept under insanitary conditions. The construction of the poultry house should receive first attention. Adequate air space, lighting, and ventilation should be provided, and the entire house should be cleaned out at frequent intervals. While these things can not be depended upon to control mites and lice, they aid the poultryman in determining when these pests are present, and, furthermore, the fowls are kept in vigorous condition, which in itself is conducive to the control of various pests. Diseased fowls, or those with malformed bills or feet, fall ready prey to lice, mites, and other insect pests. The suggestions given in previous pages for the construction of roosts and nest boxes should be followed, even though the buildings are new and otherwise properly arranged.

DUST BATHS.

While it is well to provide a good dust bath for chickens, it can not be depended upon for louse and mite control. It is far better to eradicate the pests completely. The main difficulty about depending upon dust baths for lice is that some fowls seldom dust themselves, and those which dust freely never free themselves completely. The dust bath should be kept under cover and may consist of fine road dust with coal ashes added.

A METHOD OF AVOIDING POULTRY PESTS.

It is possible for a prospective poultryman to avoid having to contend with most poultry parasites by selecting a site which is fairly well isolated from other poultry. It should be first securely fenced and new buildings and runs constructed. He should start with incubator chickens hatched on the premises and never bring any fowls on the place. Second-hand crates should not be brought on the farm unless carefully disinfected beforehand. The possibility of insects being carried from infested quarters on clothing, wagons, etc., should be kept in mind, as well as the possibility of their carriage by sparrows.

PUBLICATIONS OF THE UNITED STATES DEPARTMENT OF AGRICULTURE RELATING TO INSECTS AFFECTING THE HEALTH OF MAN AND DOMESTIC ANIMALS.

AVAILABLE FOR FREE DISTRIBUTION.

Poultry Management. (Farmers' Bulletin 287.)
Remedies and Preventives Against Mosquitoes. (Farmers' Bulletin 444.)
Some Facts about Malaria. (Farmers' Bulletin 450.)
Sanitary Privy. (Farmers' Bulletin 463.)
Hints to Poultry Raisers. (Farmers' Bulletin 528.)
Stable Fly. (Farmers' Bulletin 540.)
Harvest Mites or "Chiggers." (Farmers' Bulletin 671.)
House Flies. (Farmers' Bulletin 679.)
Hydrocyanic-acid Gas against Household Insects. (Farmers' Bulletin 699.)
Fleas as Pests of Man and Animals. (Farmers' Bulletin 683.)
Fly Traps and Their Operation. (Farmers' Bulletin 734.)
The Bedbug. (Farmers' Bulletin 754.)
Fleas. (Department Bulletin 248.)
Notes on Five North American Buffalo Gnats of the Genus Simulium. (Department Bulletin 329.)
Distribution of Rocky Mountain Spotted-fever Tick. (Entomology Circular 136.)

FOR SALE BY THE SUPERINTENDENT OF DOCUMENTS.

How to Prevent Typhoid Fever. (Farmers' Bulletin 478.) 1911. Price, 5 cents.
Yellow-fever Mosquito. (Farmers' Bulletin 547.) 1913. Price, 5 cents.
Experiments in Use of Sheep in Eradication of Rocky Mountain Spotted Fever Tick. (Department Bulletin 45.) 1913. Price, 5 cents.
Notes on the Preoviposition Period of the House Fly Musca domestica L. (Department Bulletin 345.) 1916. Price, 5 cents.
Experiments During 1915 in the Destruction of Fly Larvae in Horse Manure. (Department Bulletin 408.) 1916. Price, 5 cents.
Principal Household Insects of the United States with Chapter on Insects Affecting Dry Vegetable Foods. (Entomology Bulletin 4, n. s.) 1896. Price, 10 cents.
Notes on Mosquitoes of the United States Giving some Account of Their Structure and Biology, with Remarks on Remedies. (Entomology Bulletin 25.) 1900. Price, 10 cents.
Notes on Punkies. (Entomology Bulletin 64, pt. 3.) 1907. Price, 5 cents.
Information Concerning North American Fever Tick, with Notes on Other Species. (Entomology Bulletin 72.) 1907. Price, 15 cents.
Economic Loss to People of United States Through Insects that Carry Disease. (Entomology Bulletin 78.) 1909. Price, 10 cents.
Preventive and Remedial Work Against Mosquitoes. (Entomology Bulletin 88.) 1910. Price, 15 cents.
Rocky Mountain Spotted Fever Tick with Special Reference to Problem of Its Control in Bitter Root Valley in Montana. (Entomology Bulletin 105.) 1911. Price, 10 cents.
Life History and Bionomics of Some North American Ticks. (Entomology Bulletin 106.) 1912. Price, 30 cents.
Ox Warble. (Entomology Circular 25.) 1897. Price, 5 cents.
Horn Fly. (Entomology Circular 115.) 1910. Price, 5 cents.
Predaceous Mite Proves Noxious to Man. (Entomology Circular 118.) 1910. Price, 5 cents.
Fowl Tick. (Entomology Circular 170.) 1913. Price, 5 cents.

www.ingramcontent.com/pod-product-compliance
Lightning Source LLC
Chambersburg PA
CBHW060007230526
45472CB00008B/1992